TECHNICAL REPORT

Earthquake Insurance and Disaster Assistance

The Effect of Catastrophe Obligation Guarantees on Federal Disaster-Assistance Expenditures in California

Tom LaTourrette • *James N. Dertouzos* • *Christina E. Steiner* • *Noreen Clancy*

Sponsored by the California Earthquake Authority

RAND INSTITUTE FOR CIVIL JUSTICE

The research described in this report was sponsored by the California Earthquake Authority and was conducted by the RAND Institute for Civil Justice, a unit of the RAND Corporation.

Library of Congress Cataloging-in-Publication Data is available for this publication.
ISBN 978-0-8330-5095-3

Published 2010 by the RAND Corporation
1776 Main Street, P.O. Box 2138, Santa Monica, CA 90407-2138
1200 South Hayes Street, Arlington, VA 22202-5050
4570 Fifth Avenue, Suite 600, Pittsburgh, PA 15213-2665
RAND URL: http://www.rand.org/
To order RAND documents or to obtain additional information, contact
Distribution Services: Telephone: (310) 451-7002;
Fax: (310) 451-6915; Email: order@rand.org

Preface

In an effort to increase the availability and affordability of insurance for catastrophic events, the Catastrophe Obligation Guarantee Act (S.886/H.R.4014) was introduced in Congress in 2009. The bill would authorize the federal government to provide committed loan guarantees to qualified state catastrophe-insurance programs to reduce the programs' cost of financial capacity used to pay insurance claims after catastrophic events. Lower capacity expense can translate directly into lower premium rates and, therefore, lower-cost insurance.

One of the arguments in support of this provision is that lower-cost catastrophe insurance could reduce federal disaster-assistance expenditures because increasing the number of residents who buy the catastrophe insurance would decrease the uninsured loss in disasters. The California Earthquake Authority (CEA), a state-managed catastrophe-insurance provider that would qualify for the loan guarantees, asked the RAND Corporation, in collaboration with Risk Management Solutions, to estimate the potential magnitude of this effect for earthquakes in California. A primary objective of this analysis is to help inform federal decisionmakers in their debate about the proposed legislation. As part of the analysis, the report examines the price elasticity for earthquake insurance and relationships between earthquake-insurance coverage and loss compensation. Consequently, the report is also anticipated to be of interest to disaster-insurance suppliers, state insurance regulators, researchers, and consumers.

The RAND Institute for Civil Justice

The mission of the RAND Institute for Civil Justice (ICJ) is to improve private and public decisionmaking on civil legal issues by supplying policymakers and the public with the results of objective, empirically based, analytic research. ICJ facilitates change in the civil justice system by analyzing trends and outcomes, identifying and evaluating policy options, and bringing together representatives of different interests to debate alternative solutions to policy problems. ICJ builds on a long tradition of RAND research characterized by an interdisciplinary, empirical approach to public policy issues and rigorous standards of quality, objectivity, and independence.

ICJ research is supported by pooled grants from corporations, trade and professional associations, and individuals; by government grants and contracts; and by private foundations. ICJ disseminates its work widely to the legal, business, and research communities and to the general public. In accordance with RAND policy, all ICJ research products are subject to peer review before publication. ICJ publications do not necessarily reflect the opinions or policies of the research sponsors or of the ICJ Board of Overseers.

Information about ICJ is available online (http://www.rand.org/icj/). Inquiries about research projects should be sent to the following address:

James Dertouzos, Director
RAND Institute for Civil Justice
1776 Main Street
P.O. Box 2138
Santa Monica, CA 90407-2138
310-393-0411 x7476
Fax: 310-451-6979
James_Dertouzos@rand.org

Contents

Figure and Tables

Figure

Tables

Summary

Only about 12 percent of insured homeowners in California have earthquake insurance, which gives rise to concern about the large proportion of losses that will go uninsured in a large earthquake. Large uninsured disaster losses have significant negative impacts, not only on individuals and communities directly affected by the disaster but also on the nation as a whole in the form of postdisaster assistance from the federal government. Federal disaster-assistance spending between 1989 and 2008 exceeded 30 percent of the disaster losses, and this ratio has been increasing over time (Cummins, Suher, and Zanjani, 2010).

In an effort to increase the availability and affordability of catastrophe insurance for homeowners, newly proposed federal legislation includes a provision for committing federal guarantees for loans to qualified state disaster-insurance programs. These catastrophe obligation guarantees would support state disaster-insurance programs when they go to the private capital markets to borrow funds for claim payments following extraordinarily large disasters.

The CEA is a state-managed, largely privately funded entity that provides residential earthquake insurance that would qualify for loan guarantees under the proposed legislation. The CEA anticipates that committed federal guarantees would reduce its need for reinsurance, which would lower its expenses and allow it to charge consumers less. This would stimulate increased earthquake-insurance coverage, resulting in lower uninsured loss in an earthquake and, ultimately, reducing demand for federal disaster assistance. Thus, providing catastrophe obligation guarantees could result in a net savings to the federal government.

This analysis uses empirical and theoretical arguments to estimate the magnitude of this potential savings. Key elements of the analysis include a cross-sectional analysis to estimate the sensitivity of consumer demand for earthquake insurance to price (the price elasticity of demand); using earthquake loss-modeling simulations to estimate the relationship between residential earthquake-insurance coverage and uninsured loss in an earthquake; and performing an empirical examination of the sensitivity of demand for federal disaster assistance to uninsured residential loss. Our analysis examines two sources of disaster assistance that would be reduced by increased residential earthquake-insurance coverage: federal subsidies on low-interest disaster home loans from the Small Business Administration (SBA) and the federal individual income tax deduction for uninsured disaster losses.

Our analysis indicates that catastrophe obligation guarantees would reduce federal disaster-assistance costs by $3 million to $7 million for every $10 billion in total earthquake loss. For a simulated magnitude-7.2 earthquake on the San Francisco peninsula segment of the San Andreas Fault, the estimated federal savings would be $88 million. Although the guarantees are expected to increase consumer demand for earthquake insurance from the CEA by about 13 percent, this ultimately translates to a much smaller effect on disaster assistance. The reason that the federal savings is not more substantial is that earthquake-insurance pricing ultimately has a modest influence on the uninsured loss in an earthquake. This occurs because only a small portion of residential earthquake losses are insured to begin with (11 percent), the increase in demand for earthquake insurance in response to a price decrease is modest (price elasticity of demand = −0.44) and applies only to the CEA share of the market (61 percent), and a given increase in take-up leads to a lesser decrease in uninsured losses, because individual losses often occur in ranges that fall below deductibles.

While our analysis indicates that the federal savings under catastrophe obligation guarantees would be modest, the Congressional Budget Office (2010) estimates that the cost to the federal government of providing catastrophe obligation guarantees would also be small. A quantitative comparison of annualized costs and benefits is not possible with available data, but we estimate that benefits would exceed costs if the annual expected total loss from earthquakes in California was $7 billion or greater.

Our findings show that changes in insurance coverage would have to be dramatic to have an appreciable impact on uninsured loss and disaster assistance. This suggests that other avenues for increasing earthquake-insurance coverage, such as increased public education and marketing and offering new earthquake-insurance products that provide more-attractive options for consumers, might warrant consideration. Increasing earthquake insurance may have benefits beyond reducing federal disaster-assistance expenditures. Uncompensated disaster losses might have far-reaching and sustained economic impacts on families and communities. Examples of such indirect losses include depletion of individual savings, losses to lenders from widespread defaulting of home mortgages, local decreases in property values and property tax revenue, increased unemployment, decreased income tax revenue, and lower business investment and entrepreneurship. Few of these impacts would be compensated by disaster-assistance programs, so they would be reduced only by increased insurance coverage.

Acknowledgments

We gratefully acknowledge Daniel P. Marshall, Bob Stewart, and Glenn A. Pomeroy of the CEA; Shawna Ackerman, a principal of Pinnacle Actuarial Resources and the CEA's consulting actuary; and Michael J. Strumwasser, regulatory counsel to the CEA, for guidance, insights, and data that were essential for this study. This work was undertaken in collaboration with Risk Management Solutions (RMS), and we gratefully acknowledge data, loss-simulation analyses, and guidance provided by Randy Schlemmer, Munish Arora, Auguste Boissonnade, Don Windeler, and Michael Brill at RMS. We thank Alan Escobar of the SBA Office of Disaster Assistance for providing the SBA disaster home loan data used in the analysis. We also thank Daniel Hoople of the Congressional Budget Office for insights about the costs of catastrophe obligation guarantees. At RAND, we thank Lloyd Dixon for valuable insights about the relationship between insurance and disaster assistance and Adrian Overton for conducting geographic information system analyses used for the price-elasticity calculations. Finally, we thank George Zanjani from Georgia State University and Paul Heaton from RAND for formal peer reviews that led to valuable improvements in the report.

Abbreviations

CEA   California Earthquake Authority

FEMA   Federal Emergency Management Agency

ITV   insured-to-value ratio

RMS   Risk Management Solutions

SBA   Small Business Administration

Introduction

Maintaining a healthy catastrophe-insurance market is challenging. Compared with other insurance risks, disasters occur with low frequency, have high consequences, and result in losses that are concentrated in time and space. These characteristics create great uncertainty about risk levels and require access to large amounts of capital, risk-transfer mechanisms (commonly, reinsurance and, more recently, securitizing devices, such as catastrophe bonds), or debt capital. These and other factors lead to high insurer costs and correspondingly high premiums and deductibles for insurance policyholders. The combination of high prices, uncertain risk, and long recurrence interval between events deters those at risk from purchasing catastrophe insurance.

The market for earthquake insurance in California is illustrative of this situation: statewide, the percentage of homeowners insurance policyholders who also have earthquake insurance (referred to as the earthquake-insurance take-up) reached a high of 36 percent the year after the 1994 Northridge earthquake and has since fallen to about 12 percent (Marshall, 2009).

Low catastrophe-insurance take-up translates to large uncompensated losses in disasters, which increases demand for alternative sources of compensation, principal among them being postdisaster government assistance. Federal disaster assistance comes in many forms, including debris removal; emergency food, clothing, shelter, and medical assistance; crisis counseling; temporary housing; home and business loans; hazard-mitigation grants; unemployment assistance; tax relief; legal counseling; and public assistance to state and local governments (Bea, 2010; FEMA, 2008). Federal disaster-assistance spending is substantial: A recent analysis by Cummins, Suher, and Zanjani (2010) estimated that total federal disaster-assistance spending between 1989 and 2008 exceeded 30 percent of the disaster losses and that this ratio has been increasing over time.

The Stafford Act,[1] which sets policy for providing federal disaster assistance, specifies that "no such person, business concern, or other entity will receive such assistance with respect to any part of such loss as to which he has received financial assistance under any other program or from insurance or any other source" (FEMA, 2007, p. 18). To the extent that disaster assistance is available, it can therefore be considered a substitute for insurance. We would therefore expect increasing catastrophe-insurance take-up to decrease disaster-assistance expenditures. Indeed, one of the principal motivations for introducing the National Flood Insurance Pro-

---

[1] Pub. L. 93-288, Robert T. Stafford Disaster Relief and Emergency Assistance Act, as amended by Pub. L. 100-707, November 23, 1988.

gram in 1968 was to stanch escalating disaster-assistance costs (GAO, 2008; Dixon et al., 2006).

Based on the principle that increasing catastrophe-insurance take-up will decrease post-disaster costs, the Catastrophe Obligation Guarantee Act (S.886/H.R.4014) has been introduced to Congress for consideration. A central objective of this legislation is to decrease the cost of catastrophe insurance and increase take-up, thereby decreasing demand for postdisaster assistance. If enacted, this legislation would obligate the federal government to guarantee postevent loans from the private capital market to qualified state catastrophe-insurance programs to help pay insurance claims. Guarantees would be made available on the condition that insured losses and loss expenses exceed a particular proportion of the state program's available cash resources, which would occur only in larger disasters. In addition, maximum aggregate guarantees would be capped at $5 billion for earthquakes and $20 billion for other disasters. Catastrophe obligation guarantees have also been introduced as Title II of the Homeowners' Defense Act (H.R.2555). Federal loan guarantees would provide state insurance programs with virtually guaranteed access to capital markets at times of market pressure or distress (for example, following a disaster) and could provide interest rates more favorable than would be possible without the guarantees. This would enhance programs' debt-based, alternative funding stream for paying claims.

The California Earthquake Authority (CEA), a state-managed organization that provides about 70 percent of the residential earthquake-insurance policies sold in California, would qualify for loan guarantees under the proposed legislation. For the CEA, the promise of loan guarantees would substitute for a substantial fraction of the reinsurance it currently purchases, allowing it to lower its reinsurance expense. Because the CEA is a tax-exempt, non–profit-earning entity and because insurance pricing is highly regulated, we assume that the savings from reducing reinsurance requirements would be passed on to policyholders in the form of reduced premiums and deductibles and improved coverage choices. The CEA anticipates that decreasing policyholder prices would stimulate increased policyholder take-up, which would then decrease uninsured losses and, ultimately, decrease federal disaster-assistance expenditures. The logical progression of this argument is summarized in Figure 1.

While the logic expressed in Figure 1 is well grounded in economic theory, the magnitude of the effect in each step is not known. The objective of this work is to estimate the change in federal disaster-assistance expenditures for earthquakes in California that would result from catastrophe obligation guarantees. We approach this by simulating the steps in the logical chain shown in Figure 1. It is important to recognize that the empirical basis for quantifying some of the effects illustrated in Figure 1 is quite limited. Consequently, considerable uncertainty remains about the magnitude of estimated effects. Further, we have not considered a range of additional costs or benefits that could affect the efficacy of increased insurance coverage for homeowners. For example, uncompensated losses due to unanticipated natural disasters could well have far-reaching and sustained economic impacts on families and the communities in which they live. We have also not considered potentially important incentive effects, such as changes in loss mitigation associated with insurance coverage. Thus, although our analysis provides useful evidence for a more-comprehensive policy analysis, additional information would be necessary in advance of more-definitive conclusions.

**Figure 1
Anticipated Effect of Catastrophe Obligation
Guarantees on Disaster Assistance**

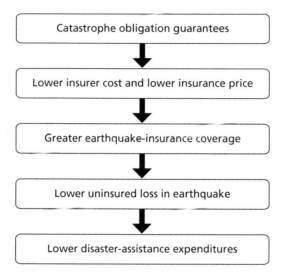

RAND *TR896-1*

Analysis Approach

In this section, we step through each of the relationships illustrated in Figure 1. Each relationship can, in principle, be modeled according to knowledge gained from past experience for similar conditions. In practice, however, the historical information needed to evaluate these relationships is generally poor. The reasons better data are not available are multifold and include the rarity of earthquakes, the absence of good loss estimates for most earthquakes that have occurred, the complex array of factors driving consumers' demand for earthquake insurance, and ambiguities in the extent to which disaster-assistance expenditures are sensitive to insurance coverage. We address each of these issues in this section and attempt to highlight significant uncertainties and areas in which having better data would improve the analysis.

Effect of Catastrophe Obligation Guarantees on Earthquake-Insurance Price

The relationship between catastrophe obligation guarantees and the price of residential earthquake insurance has been modeled by the CEA, and we adopt this estimate in our analysis. Based on models of expected earthquake losses, the probability of needing to utilize catastrophe obligation guarantees, and the cost of reinsurance that could be discontinued, the CEA concluded that policyholder prices for its customers are expected to decrease by approximately 30 percent with the catastrophe obligation guarantees in place (Pomeroy, 2010). This decrease is not expected to depend on location or risk, so it is presumed to be uniform throughout the state. Alternatively, the CEA could reduce deductibles, which would also serve to increase the amount of property insured against earthquake risk. The amount by which deductibles could be decreased under catastrophe obligation guarantees is not known, and there is scant information available on which to base an estimate of how a change in deductibles might influence take-up. Hence, we are not able to examine the trade-off between decreased premiums and decreased deductibles that would be possible under catastrophe obligation guarantees. This topic is worthy of future research.

Effect of Price on Amount of Earthquake-Insurance Coverage

The relationship between price and demand for earthquake insurance can be expressed as the price elasticity of demand. The price elasticity of demand is defined as the fractional change in demand divided by the fractional change in price. For most products and services, the amount demanded increases when the price decreases. In other words, the price elasticity is negative.

In the case of catastrophe insurance, demand is typically discussed in terms of policy-holder take-up, which is the fraction of consumers already insured under a policy of residential property insurance who also purchase separate catastrophe insurance. For evaluating the impact of uninsured losses on subsequent federal assistance, however, the fraction of residential property value that is insured for earthquake loss, or the *insured-to-value* ratio (ITV), might be more relevant than the take-up. We define the ITV as the total structural coverage (sum of dwelling [*coverage A*] policy limits minus sum of policy deductibles) for earthquake insurance divided by the total residential housing structural value.[1] This is a more-useful metric for our purposes because we are ultimately interested in estimating changes in the amount of uninsured loss, which depends on how much property loss is reimbursed by earthquake insurance. Earthquake insurance typically has a much larger deductible (generally, 15 percent of the limit of structure coverage) than homeowners insurance has. Also, while policy limits for CEA earthquake insurance are legally required to match limits for general homeowners insurance, these limits are not necessarily as high as the actual structure value. Consequently, the actual fraction of property value that would be reimbursed through earthquake insurance might be less than the take-up.

Another reason for using ITV rather than take-up is that ITV can increase both by uninsured consumers purchasing new insurance and by insured consumers decreasing their deductibles. ITV captures both of these effects, while take-up would capture only the first and miss any increase in coverage under existing policies.

The current ITV of earthquake insurance in California was estimated from data on earthquake-insurance coverage and residential structural property value in California. In December 2009, the sum of coverage A provided by the CEA was $247 billion, and the sum of deductibles on CEA policies was $35.9 billion. The difference between these values, $211 billion, is the total structural earthquake-insurance coverage provided by the CEA. Data from the California Department of Insurance (2010) indicate that the CEA had 61 percent of the total exposure on all earthquake coverage in 2009. No data on the relative proportions of separate coverage types (i.e., structure, personal property, additional living expense) for non-CEA earthquake insurance were available. If we assume that the 61-percent market share for total CEA coverage applies to structural coverage as well, then we can infer that the overall residential structural earthquake coverage is $1/0.61 = 1.6$ times greater than the CEA coverage, or $347 billion. Risk Management Solutions (RMS) maintains a database of residential structural value in California. In December 2009, this value was $3,183 billion. Dividing the overall residential structural earthquake coverage by this structural value gives a current ITV of 10.9 percent.

A decrease in the price of earthquake insurance under the catastrophe obligation guarantee will translate to an increase in ITV by means of the relationship (see appendix)

$$ITV_{COG} = ITV_o \left(1 + 0.61 E \left(\frac{\Delta p}{p_o} \right)_{CEA} \right).$$

(1)

---

[1]   Note that this definition differs from the conventional definition of ITV in that it excludes the deductible amount.

The subscripts *COG* and *o* indicate conditions with and without the catastrophe obligation guarantees, respectively. E is the price elasticity of demand, and $\left(\Delta p/p_o\right)_{CEA}$ is the fractional price change of CEA insurance with catastrophe obligation guarantees. Because only state catastrophe-insurance programs are eligible for catastrophe obligation guarantees, any resulting price change would apply to CEA policies only. The factor of 0.61 appears because the CEA writes 61 percent of the earthquake-insurance coverage in California.

Price Elasticity of Demand

The literature on price elasticity for catastrophe insurance is sparse. Only a few estimates of price elasticity are available for flood insurance, and none has been reported for earthquake insurance. Estimates for flood insurance range from –0.39 to –0.997 (GAO, 1983; Browne and Hoyt, 2000; Kriesel and Landry, 2004; Hung, 2009).[2] Given the great differences between risks and insurance options for floods and earthquakes, however, elasticity estimates for flood insurance might have limited relevance to earthquake insurance. We therefore developed our own estimate of the price elasticity for earthquake-insurance demand.

We examined the relationship between premium rates and the number of households purchasing CEA coverage during the fourth quarter of 2005. Cross-section information was gathered for 1,070 California ZIP Codes,[3] indicating the number of CEA policies sold, the total number of homeowners with residential insurance policies, premium rates for a standardized policy,[4] and several covariates likely to affect the level of demand, including local demographics, household income, and housing values. The data set is described in Table 1.

The identification of a pure demand relationship between price and quantity is challenging. This is because CEA premium rate levels are determined in 19 rating territories based on an assessment of the average liability risk in that geographic region. Since the value of an earthquake policy is greater in high-risk areas, households will be willing to pay more to purchase a policy, all other things equal.

To frame the issue, assume the following:

Q_i = number of people taking insurance in ZIP Code i
P_i = price of insurance
$\overline{R}_i$ = average true risk in a rating territory
R_i = perceived risk for a household
X_i = exogenous factors, such as income and demographics
r_i = true risk factors, such as proximity to fault line or high-shake territory
P_j = price of insurance in a contiguous rating territory.

---

[2] We are aware of the one exception in the literature (Grace, Klein, and Kleindorfer, 2004) that estimates a significantly higher demand elasticity for residential insurance, but, since that study bundles the prices of standard residential coverage and catastrophe insurance, the computed elasticities are not comparable.

[3] The analysis file includes only those California ZIP Codes for which there were census data describing market characteristics. Thus, small areas including few numbers of households were excluded. The remaining sample represents more than 94 percent of California's residential homeowners.

[4] The standardized policy was for a home value of $250,000, which is the typical structure coverage of a CEA policy. The choice of a $250,000 benchmark is not relevant to the analysis, since premium rates per dollar of coverage are linear across levels of housing value. Thus, the computed elasticity would be identical for any alternative benchmarks.

Table 1
Data Used in Analysis of Earthquake-Insurance Take-Up

Variable	Mean	Standard Deviation
Total residential home policies	3,315	4,181
Earthquake policies	336	490
Average annual premium (for $250,000 in coverage) ($)	271	176
High-risk shake (0,1)	0.410	0.492
High-risk fault (0,1)	0.244	0.430
Average home value ($)	146,729	179,949
Income per household ($)	30,032	28,372
Median age of residents (years)	23.6	18.7
Business payroll	235,941	597,074
Percentage African American	0.146	0.732
Percentage Hispanic	0.519	0.713
Land area (square miles)	43.6	115.7
Multifamily dwellings	1,099	2,416
Elevation above sea level	442.0	799.8
Premium price (relative to bordering ZIP)	1.086	0.564

Demand for insurance will be based on price, the perceived risk of an earthquake, and a set of exogenous factors. In other words,

$$Q_i = f\left(P_i, R_i, X_i\right).\tag{2}$$

At the same time, price is determined by the average risk in a rating territory. In other words,

$$P_i = g\left(\overline{R_i}\right).\tag{3}$$

If perceived risk in individual ZIP Codes deviates from the average risk in a territory in a systematic and measurable manner, then it becomes possible to identify the separate impact of price. For example, imagine that the perceived risk within a territory is given by

$$R_i = h\left(r_i\right).\tag{4}$$

It should be possible to implement a strategy that takes advantage of variations in perceived risk or price. In the analysis that follows, we control for perceived risk in two ways. First, we allow for the possibility that location in a high-risk shake zone or near a fault line increases perception of risk. Next, we examine contiguous ZIPs located in different rating territories, and we assume that contiguous geographic areas have similar perceived risks.[5]

To demonstrate the value of the first strategy, Table 2 provides the percentage of ZIP Codes that are determined to be in a high-risk shake zone or close to a fault line within each of the 19 rating territories. To the extent that these risk-indicator variables signal within-territory risk differences, they should affect willingness to purchase insurance at the prevailing premium price. For example, in rating territory 5, the standard premium for a $250,000 policy would be $743. However, note that only 69 percent of the ZIP Codes are located in high-risk fault areas. Thus, one would expect that the other 31 percent would value the coverage somewhat less, thereby reducing demand for coverage at the prevailing price. Within-territory variation in these risk measures, despite homogeneous pricing of policies, provides an opportunity to identify the impact of a risk-neutral price difference.

Table 2
Variations in Risk Within Rating Territories

Rating Territory	High-Risk Shake ZIP Code (%)	ZIP Code on Fault Line (%)	Standardized Premium ($)
2	100	92	490
4	100	100	743
5	100	69	743
6	90	39	463
7	99	40	318
8	100	64	648
11	92	71	468
12	98	46	458
13	100	67	295
15	85	73	245
18	97	57	90
19	100	40	225
20	97	74	335
22	97	49	578
23	100	61	443
24	99	46	328
25	100	96	435
26	94	57	353
27	35	30	105

---

[5] It is important to note that, if our controls for variations in perceived risk are not adequate, then the relationship between insurance premiums and take-up will combine both a pure price effect and a response to perceived risk. If this is the case, then the estimated elasticity must be viewed as a conservative or lower-bound estimate of the true elasticity.

Of course, the CEA's share of the California earthquake-insurance market is about 61 percent. Due to the absence of relevant data, the regressions do not control for the pricing behavior of private competitors and, therefore, the impact of CEA pricing on the volume of business earned by other insurance companies. This will not bias the elasticity estimate, but the relationship between take-up and premium levels does not account for the level of total earthquake coverage across the state. Rather, the expansion effect measures the CEA portion of the market only.[6]

To implement this strategy for identifying the impact of price on earthquake coverage, a statistical model was estimated, linking the number of CEA policies with a set of independent variables, including the number of residential home policies, premium, risk indicators, and several covariates. The results of this model estimation are reported in Table 3.[7]

When one controls for other factors, including indicators likely to be correlated with the within-zone variation in perceived risk, premium price appears to have a significant and negative effect on demand. Since the model is specified as a log-linear relationship, the estimated coefficient of -0.4814 represents the price elasticity of demand. In other words, a 10-percent decline in price would result in about a 4.8-percent increase in the number of households purchasing earthquake insurance.[8] This relationship is based on holding other factors constant, including the level of perceived risk, household income, housing values, and demographics, such as age, race, and other ZIP Code characteristics.

It is worth noting that the estimated impacts of other covariates are generally consistent with economic theory and intuition. For example, the number of earthquake policies expands roughly in proportion with the total number of residential home policies. When prices are held constant, the number of CEA policies increases significantly when the ZIP Code is a designated high-risk shake area. This reflects the fact that the territory's premium reflects the average risk across all individual ZIP Codes. The impact of being in a high-risk fault area is positive but much less important.

As one would expect, demand for insurance increases with household income and housing values.[9] Hispanic populations appear to be less likely to purchase insurance, while the opposite is true of African American populations. Older populations are more likely to purchase insurance. Other control variables that appear to be significant include the elevation, land area, business activity, and the presence of multifamily dwellings.

---

[6]   There is no reliable evidence about whether or not households or their insurance brokers actively shop for lower coverage premiums across competitors. In the absence of evidence about cross-price elasticities, we can do no more than acknowledge that some portion of the expanded take-up rate induced by lower CEA premiums could include customers who merely switched carriers. This would tend to diminish the overall expansion of coverage.

[7]   We also explored weighted least squares regressions that account for the possibility of residual correlations within territories. In fact, about 30 percent of the variation in residuals could be explained by a set of 19 dichotomous variables representing the different pricing territories. However, in regressions that netted out, these differences were not materially different from our basic ordinary least squares model.

[8]   With an estimated coefficient that is about 12 times the magnitude of the standard error, one can state with 99.9 percent confidence that the impact of a price reduction on take-up is greater than 0. Most of the other variables have coefficient/standard error ratios of greater than 2.0, indicating confidence of at least 95 percent. The only exceptions are the high-risk fault areas and the local business payroll variables, which are slightly less significant at the 90-percent level of confidence.

[9]   Although the probability of insurance take-up increases with housing value, the total insured value in a community does not increase proportionately. This is because of the increased importance of land value in wealthier communities, as well as limits on policy coverage. As a result, regressions based on total insured value rather than take-up rates had virtually identical results.

Table 3
The Demand for Earthquake Insurance: Controlling for Variations in Risk

Variable	Coefficient	Standard Error
Intercept	−14.1204	0.9883
Log(residential home policies)	1.0632	0.0298
Log(premium price)	−0.4814	0.0423
High-risk shake ZIP	1.3162	0.0677
High-risk fault ZIP	0.0744	0.0437
Log(household income)	0.1605	0.0929
Log(average housing value)	0.6211	0.0706
Log(percentage African American)	0.0371	0.0172
Log(land area)	−0.0482	0.0174
Log(multiple-family dwellings)	0.0883	0.0195
Log(elevation)	0.1390	0.0158
Log(percentage Hispanic)	−0.1564	0.0331
Log(median age)	0.6099	0.1845
Log(business payroll)	−0.0350	0.0186
R-squared	0.8167	

NOTE: Dependent variable = log(CEA earthquake policies). N = 1,070.

We also explored whether total insured value increased at the same rate as the take-up rate. We ran regressions with total insured value, and the elasticity was virtually identical, at −0.501. In addition, models that examined supplemental coverage were examined. The results indicated that average coverage for personal property losses and transitional expenses were not affected by variations in price. This indicates that the estimated impact on take-up rates is an accurate gauge of the overall effect on the percentage of property value covered by insurance.

The estimated price elasticity is accurate if and only if the included covariates reflect interterritory variations in perceived earthquake risk. As we have seen, the indicator variable representing shake risk is quite significant in explaining ZIP Code variations in demand for earthquake insurance. However, as a reliability test, we identified all those ZIP Codes that bordered ZIPs located in different pricing territories. Although such contiguous ZIPs are subject to different pricing schemes based on the average risk in their territories, they are likely to be subject to similar earthquake risks due to their geographic proximity. For contiguous areas, we therefore assumed that their risk was similar and that the premium differences represent pure price differences, holding risk constant.[10]

In this analysis, we regressed the difference in take-up rates (CEA policies/residential insurance policies) in contiguous areas on differences in the logarithm of price and differences in the set of covariates described in Table 1. For some of the variables (such as the shake-risk

[10] There were a total of 325 ZIP Codes that bordered on other pricing territories. In many cases, a ZIP bordered on multiple ZIP Codes located in contiguous pricing territories. In these cases, the bordering ZIPs were combined.

Table 4
The Demand for Earthquake Insurance: Comparing Outcomes from Territory Borders

Variable	Coefficient	Standard Error
Log difference (premium price)	−0.3952	0.1445
Log difference (household income)	0.1743	0.2095
Log difference (housing value)	0.2746	0.2121
Log difference (% African American)	0.1945	0.0392
Log difference (land area)	−0.0119	0.0224
Log difference (multiple-family dwellings)	−0.0170	0.0106
Log difference (elevation)	−0.0404	0.0529
Log difference (% Hispanic)	−0.5516	0.0597
Log difference (median age)	−0.1894	0.2971
Log difference (business payroll)	−0.0213	0.0233
R-squared	0.4060	

NOTE: Dependent variable: log(difference in take-up rate). N = 325.

indicator), the variation in values for contiguous ZIP Codes was insufficient for estimating their impact. However, the variation for premium rates based on territory was significant and, as we can see in Table 4, was sufficient to estimate an impact on take-up rates.

In particular, the estimated price elasticity was −0.3952, a result that is not significantly different from the previous elasticity estimate. This result strengthens the confidence that we have isolated the independent effect of price on demand for earthquake insurance and controlled for the role of liability risk in the determination of premiums. Although still somewhat inelastic, it is clear that a decline in premium pricing will result in a significant increase in insurance coverage.

The average of our two elasticity estimates is −0.44. For the anticipated −30-percent price change under catastrophe obligation guarantees, our analysis indicates that the take-up of CEA earthquake insurance would increase by 13.2 percent. Because our separate estimates that examined take-up and the total amount of earthquake coverage gave virtually identical elasticities, our elasticity estimate applies to both take-up and ITV. Proprietary survey findings obtained by the CEA are highly consistent with our results.

Effect of Policyholder Take-Up on Uninsured Loss

The next step in our analysis is to determine how a change in ITV would influence the uninsured loss in an earthquake. Increasing ITV will shift some of the loss from being uninsured to being insured, which will decrease the uninsured loss. Because the CEA provides residential property insurance, our analysis is restricted to the change in uninsured residential loss. Catastrophe obligation guarantees extended to the CEA will not influence commercial, medical-, or life-insurance losses in California.

The relationship between ITV and the fraction of earthquake loss that is reimbursed by insurance is complicated by the existence of policy deductibles. This complexity can be illustrated by considering two hypothetical, end-member outcomes for the same ITV and total earthquake loss. In the first outcome, the total loss is distributed among many properties such that the loss to each individual property is less than the policy deductible. In this case, none of the loss is reimbursed by insurance. In the second outcome, the total loss is concentrated into a small number of properties such that each property suffers a total loss. In this case, all of the loss above the deductibles is reimbursed, and the fraction of loss reimbursed by insurance equals the ITV. The fraction of loss reimbursed by insurance therefore depends on the ratio of loss to value among residential properties damaged in an earthquake. This ratio will vary among damaged properties, and the distribution of this ratio among properties depends on the details of the earthquake scenario and is difficult to generalize.

We estimated the fraction of loss reimbursed by insurance for a range of conditions using earthquake loss-modeling simulations conducted by RMS. The fraction of residential loss reimbursed by insurance is given by L_R^I/L_R, where L_R^I is the residential loss reimbursed by insurance and L_R is the total residential loss (insured plus uninsured). The simulations give an estimate of L_R^I/L_R, as a function of the fraction of residential property value insured, or the ITV. That is, they provide the value of R_I in the relationship

$$L_R^I/L_R = R_I ITV.$$

(5)

Simulations were run for the probability-weighted sum of all earthquake events expected in California, as well as for subsets of earthquakes representing increasingly large (and less probable) events. These subsets are characterized by the annual probability of occurrence and the associated expected recurrence interval. Results of the simulations are shown in Table 5.

The simulation results show that R_I ranges from a minimum of about 0.3 for the smallest earthquakes to a maximum of about 0.7 for very rare large earthquakes. For the current ITV of 10.9 percent, a value of $R_I = 0.34$ means that only 3.7 percent of the residential losses in an

Table 5
Relationship Between Fraction of Earthquake Loss Reimbursed by Insurance and Insured-to-Value for Different Conditions

Annual Probability of Occurrence	Economic Recurrence Interval	$R_I = \dfrac{\left(I_R^I/L_R\right)}{ITV}$
<100% (all events)	>1 year	0.34
<50%	>2 years	0.44
<5%	>20 years	0.57
<1%	>100 years	0.63
<0.5%	>200 years	0.66
<0.2%	>500 years	0.68
<0.1%	>1,000 years	0.73

SOURCE: RMS simulations conducted for this analysis.

earthquake would actually be reimbursed by insurance. The difference reflects losses to insured properties that would fall below deductibles and not be reimbursed. As the average size of the earthquakes being considered increases, the proportion of properties suffering more-extensive damage increases, resulting in an increase in the fraction of loss reimbursed by insurance. This increase is not linear, however, and R_l plateaus at about 0.7. Considering only larger, less frequent earthquakes, the fraction of loss reimbursed by insurance could be as high as 8 percent.

Equation 5 can be rewritten in terms of uninsured residential loss, L_R^U, to give (see appendix)

$$L_R^U = L_R \left(1 - R_l ITV\right). \qquad (6)$$

Finally, we can recast this in terms of total earthquake loss by defining R_r as the ratio of the total residential loss to total loss, L_R/L_T:

$$L_R^U = R_r L_T \left(1 - R_l ITV\right). \qquad (7)$$

The value of R_r must be estimated or determined empirically from representative earthquake scenarios. We use two loss-distribution estimates as the basis for our modeling. The first is from the 1994 Northridge earthquake, and the second is for a simulation of a magnitude-7.2 earthquake on the San Francisco peninsula segment of the San Andreas Fault. Loss distributions for these two scenarios are shown in Table 6.

Aside from the differing total loss amounts, the major difference between the two scenarios is the fraction of residential loss that is insured. Most of this difference reflects the decrease in residential earthquake-insurance take-up and standard policy coverage between 1994 and 2009. The distributions are otherwise quite similar. In particular, the ratio of residential loss to total loss, R_r, is very close in both cases (last column). Although this ratio could vary in small earthquakes due to idiosyncrasies in the local built environment, it will converge toward the

Table 6
Loss Distributions for Two Representative Earthquake Scenarios

Scenario	In Billions of Dollars			As Percentage of Total		
	Insured	Uninsured	Total	Insured	Uninsured	Total
1994 Northridge quake						
Residential	9.88	10.7	20.6	23.6	25.6	49.3
Nonresidential	4.02	17.2	21.2	9.62	41.1	50.7
Total	13.9	27.9	41.8	33.3	66.7	100
2009 San Andreas simulation						
Residential	4.1	51	55.1	3.45	42.9	46.3
Nonresidential	14.7	49.1	63.8	12.4	41.3	53.7
Total	18.8	100	119	15.8	84.2	100

SOURCES: Northridge earthquake, Petak and Elahi (2000). San Andreas simulation, Grossi and Zoback (2009).

values in Table 6 for large earthquakes in western California that span substantial geographic areas. We use the average value ($R_r = 0.48$) in our analysis.

Effect of Uninsured Loss on Disaster Assistance

The final step in our analysis is estimating how a change in uninsured loss translates into a change in federal disaster-assistance expenditures. To a first-order approximation, federal disaster assistance is intended to compensate individuals and communities for disaster losses that are not compensated for by some other mechanism. Since insurance is the principal compensation mechanism for disaster losses, we expect the amount of disaster assistance to increase with the fraction of loss that is uninsured.

To quantify this effect, we would ideally examine disaster assistance for several earthquakes of similar total loss but with widely varying uninsured loss. However, such an approach is complicated by several factors. First, earthquakes are rare events—there have been only seven presidentially declared earthquake disasters in California since 1989. Second, estimates of total loss (and, by difference, uninsured loss) are quite difficult to compile and are rarely reported. While insured disaster losses are recorded in insurance claims, and aggregated results are reported by state insurance departments, the concept of total loss from a disaster is less clearly defined, and no entity keeps track of total losses resulting from catastrophes in the United States (Lott and Ross, 2006). Of the seven earthquake disasters since 1989, an estimate of the uninsured residential loss is available only for the 1994 Northridge earthquake. Third, there is tremendous variation in the extent to which disaster-assistance expenditures are sensitive to uninsured loss in general, and to uninsured residential loss in particular.[11] This means that the change in total disaster assistance resulting from a change in uninsured residential loss will be small and obscured by the noise in overall disaster-assistance data.

In light of these limitations, we focused on two principal sources of disaster assistance available for uninsured property losses: federal income tax deductions for disaster losses and Small Business Administration (SBA) disaster home loans.

Federal Income Tax Deduction for Disaster Losses

The federal tax code allows taxpayers to claim unreimbursed casualty, disaster, and theft losses as a deduction on their federal personal income tax (IRS, 2009). Although not a direct expenditure by the federal government, a tax deduction is nonetheless an important form of federal disaster assistance. For example, special legislation passed in the wake of specific disasters often includes personal income tax relief (e.g., the Katrina Emergency Tax Relief Act[12] and Gulf Opportunity Zone Act of 2005[13] [IRS, 2007] and the Victims of Terrorism Tax Relief Act of 2001[14] [Wolfe, 2003]).

---

[11] Examples of disaster assistance that does not substitute for uninsured residential property loss include public assistance, business grants and loans, and hazard-mitigation grants.

[12] Pub. L. 109-73.

[13] Pub. L. 109-135.

[14] Pub. L. 107-134.

This deduction can be claimed as an adjustment to the standard deduction or as an itemized deduction, so it is available to all taxpayers. If we assume that all homeowners file tax returns, then this deduction can be applied to all uninsured residential loss. Tax deductions work by reducing the amount of income on which tax must be paid; the value of a tax deduction is therefore the amount of the deduction multiplied by the marginal tax rate. The average marginal federal individual income tax rate in the United States from 2003 to 2006 (the most recent year for which data are available) was 21.4 percent (Barro and Redlick, 2009). We can therefore compute the federal assistance from the disaster loss tax deduction, F_{tax}, from

$$F_{tax} = 0.214 L_R^U.$$

(8)

The average marginal income tax rate in 1994 was 23 percent. Thus, for the $10.7 billion in uninsured residential loss from the Northridge earthquake (Table 6), the federal assistance available from the disaster loss tax deduction was $2.5 billion.

Small Business Administration Loans

Low-interest SBA disaster home loans are the primary source of direct government assistance for residential property owners and are designed specifically for those without insurance and those not fully covered by insurance (FEMA, 2008, 2009). Even after accounting for the fact that most loans are repaid, the final federal expenditure after loan defaults is a major source of federal disaster-assistance spending for residential losses.

Unlike a tax deduction, the amount of federal spending through SBA loans cannot be computed and must be determined empirically. We obtained data from SBA on individual disaster home loan amounts and annual subsidy rates for each of the seven presidentially declared earthquake disasters in California since 1989 (99,700 loans). Linking SBA loan amounts to uninsured residential loss requires estimates for total and uninsured earthquake losses, which, as noted above, are available only for the Northridge earthquake. Approximately 83 percent of the loans and 83 percent of the total amount loaned in our data set was for the Northridge earthquake. Hence, although our analysis is restricted to a single event, that event comprises the vast majority of all earthquake loans in the past 20 years.

To accurately model the federal expenditures through the SBA disaster home loan program that would be expected to vary with residential earthquake insurance, the SBA loan data must be adjusted in two ways. First, because we are interested in the amount of federal assistance that would be replaced by earthquake insurance, we consider only that portion of each loan that exceeds the average earthquake-insurance policy deductible at the time of the Northridge earthquake. This is because any loan or portion of a loan less than the deductible amount would have been made whether or not the applicant had insurance and, hence, would still be made even if everyone had insurance. Eguchi et al. (1998) estimate that the average earthquake-policy deductible in the Northridge earthquake was $15,000. Of the $1.77 billion in SBA disaster home loans, $837 million (47 percent) was for amounts above the $15,000 deductible.

Second, because we are interested federal expenditures (rather than the amount lent), we need to know the government cost of providing SBA loans. The SBA reports this in the form of an annual disaster loan subsidy rate, which is the difference between the net present value of expected cash flows and the face value of a loan. This takes into consideration such factors as the expected repayments on the loans, the differential between the interest rate for SBA-held loans and the rate at which funds are borrowed by the Treasury, and the change in the value of money over the lifetime of the loan. The federal expenditure for SBA loans is the loan amount multiplied by the subsidy rate. SBA made loans for the Northridge earthquake in 1994, 1995, and 1996, so we used the average subsidy rate for these three years (27 percent).

The final federal expenditure for SBA loans that substituted for insurance in the Northridge earthquake was therefore 27 percent of $837 million, or $226 million. By normalizing this quantity to the uninsured residential loss in the Northridge earthquake ($10.7 billion, Table 6), we derive a constant of proportionality that can be used to estimate the SBA disaster home loan expenditures that would be replaced by insurance in any earthquake: $R_s = 226/1{,}070 = 0.021$. The federal expenditure for SBA loans that substitute for insurance, F_{SBA}, can then be expressed as

$$F_{SBA} = R_s L_R^U.$$
(9)

F_{SBA} is not the total SBA expenditures but rather only that portion of those expenditures that substitute for insurance. It is important to emphasize that Equation 9 is calibrated only to the Northridge earthquake. The eligibility criteria for SBA disaster home loans clearly indicate that loans are available only for uninsured losses, and we are therefore reasonably confident that SBA expenditures are proportional to uninsured residential loss. It is possible, however, that the constant of proportionality, R_s, has changed since 1994. For example, residential property values and home repair costs have increased since 1994, but the maximum limit on SBA disaster home loans has not. If SBA loan amounts have not kept pace with uninsured earthquake losses, this might suggest that R_s is lower today than it was in 1994. However, 98 percent of the SBA earthquake loans since 1989, comprising 80 percent of the total amount loaned, were for less than $100,000, which is less than half the maximum SBA loan amount of $240,000 (FEMA, 2009). This finding continues to hold when considering only loans since 2005. This indicates that SBA expenditures are not constrained by the maximum loan limit and, hence, that R_s will not have changed for this reason.

Another major change since 1994 is that residential earthquake-insurance policy deductibles in California have increased by a factor of three, to an average of about $45,000. This increase results from both increased property values and an increase in the typical deductible percentage from 10 percent to 15 percent. Uninsured loss includes both losses to uninsured property and deductibles on insured property so that, with everything else held constant, higher deductibles would increase the uninsured residential loss in an earthquake. However, greater uninsured loss would increase demand for disaster assistance, so greater deductibles would presumably be associated with concomitantly greater SBA loan expenditures. Variation in deductibles is therefore unlikely to have any effect on R_s.

Other reasons that R_s might change from earthquake to earthquake, such as variations in the availability of alternative forms of disaster assistance or people's willingness to apply for assistance, cannot be ruled out but cannot be examined without data on additional earthquakes.

Total Federal Assistance for Uninsured Residential Loss

Summing Equations 8 and 9 gives a total federal expenditure of

$$F_{Total} = \left(0.214 + R_s\right)L_R^U.$$ (10)

Our analysis has focused on income tax deductions and SBA loans, but additional sources of disaster assistance to residential property owners is available, most notably through the Federal Emergency Management Agency (FEMA) Individuals and Households Program. To the extent that this assistance also substitutes for insurance, our estimate of the effect of increased insurance coverage on disaster assistance is a conservative, lower-bound value. Although we were not able to obtain data on FEMA disaster-assistance expenditures, two observations suggest that the inclusion of FEMA assistance would have a small effect on our results. First, total FEMA assistance is limited to $28,800 (FEMA, 2009, Annex 7), which is less than the standard 15-percent deductible for an earthquake-insurance policy of $192,000 or more. Given home values in earthquake-prone regions in California, this limit suggests that most FEMA assistance would be below deductibles and therefore would not be reduced by an increase in insurance coverage. Second, all applicants for long-term disaster housing assistance are first referred to SBA for assistance; only those applicants with an unacceptable credit history or insufficient income receive disaster housing assistance from FEMA. Property owners receiving FEMA assistance might therefore be less likely than those receiving SBA assistance to be in a position to purchase insurance if prices decrease. This again suggests that the bulk of FEMA disaster assistance might not be affected by an increase in insurance coverage.

Net Effect of Catastrophe Obligation Guarantees on Disaster Assistance

Combining Equations 7 and 10 gives the relationship between the ITV for earthquake insurance and federal disaster assistance:

$$F_{Total} = \left(0.214 + R_s\right)R_r L_T \left(1 - R_I ITV\right).$$ (11)

For a given ITV, Equation 11 gives the amount of federal disaster assistance that would be replaced by insurance as a function of total earthquake loss. The difference in disaster assistance with and without catastrophe obligation guarantees can be determined by evaluating Equation 11 at the current ITV and at the expected ITV with catastrophe obligation guarantees (given by Equation 1).

Results

Federal Savings from Catastrophe Obligation Guarantees in California

Our analysis allows us to estimate the effect of earthquake-insurance pricing on uninsured residential loss and federal disaster-assistance expenditures. The results are expressed in terms of six key parameters: the fractional price change in CEA policies, the current ITV for residential earthquake insurance, the price elasticity of demand for residential earthquake insurance, and three ratios—the ratio of the fraction of loss reimbursed by insurance to the ITV (R_l), the ratio of residential loss to total loss in an earthquake (R_r), and the ratio of SBA disaster home loan expenditures that substitute for insurance to the uninsured residential loss (R_s). Values for these parameters are summarized in Table 7.

While there is uncertainty associated with all of the parameters in Table 7, we focus our examination of uncertainty on R_l because it is very sensitive to the earthquake scenario and hence has the greatest natural variation. While R_s might also vary from earthquake to earth-

Table 7
Key Parameters Linking California Earthquake Authority Residential Earthquake-Insurance Price to Disaster Assistance

Parameter	Estimate	Source
$\Delta p / p_o$, fractional price change of CEA insurance	−0.3	CEA
ITV_o, current ITV for residential earthquake insurance in California	0.109	Our analysis of CEA and RMS data
E, price elasticity of demand	−0.44	Our analysis of CEA data
$R_l = \dfrac{\left(L_R^I / L_R\right)}{ITV}$	All earthquakes: 0.34 Large earthquakes: 0.73	RMS simulation results
$R_r = \dfrac{L_R}{L_T}$	0.48	Table 6
$R_s = \left(\dfrac{SBA\ cost\ above\ deductibles}{L_R^U}\right)_{Northridge}$	0.021	Our analysis of SBA data and Table 6

quake, it ultimately has a small influence on the results because federal expenditures are dominated by the disaster loss tax deduction.

Results are presented in Table 8 in terms of the change that would occur with catastrophe obligation guarantees for every $10 billion in total earthquake loss. Results are shown for two sets of conditions that span the range in the net impact of catastrophe obligation guarantees on federal disaster assistance. The first case represents conditions that hold when considering all earthquakes. As such, it approximates the time-integrated average, or steady-state conditions. The majority of earthquakes are small, resulting in a substantial fraction of residential loss falling below policy deductibles and going uninsured. This, in turn, leads to a small decrease in uninsured residential loss for a given increase in ITV. This ultimately translates to about a $3 million decrease in federal disaster assistance. The second case represents conditions that hold only for very large earthquakes, in which a greater fraction of residential loss exceeds policy deductibles and hence a smaller fraction of residential loss goes uninsured. This leads to a greater decrease in uninsured residential loss with increasing ITV. The $7 million decrease in federal disaster assistance thus represents an upper bound on the impact of catastrophe obligation guarantees on disaster assistance.

Our results show that the ultimate federal savings from catastrophe obligation guarantees is expected to be modest ($3 million to $7 million for every $10 billion in total earthquake loss). For the simulated San Andreas Fault earthquake with a total loss of $119 billion shown in Table 6, the estimated federal savings using the R_l for large earthquakes would be $88 million. The reason that the federal savings is not more substantial is that earthquake-insurance pricing ultimately has only a modest influence on the amount of uninsured residential loss in an earthquake. This occurs because only a small portion of residential earthquake losses are insured to begin with, the increase in demand for earthquake insurance in response to a price decrease is modest and applies only to the CEA share of the market, and a given increase in take-up or ITV leads to a lesser decrease in uninsured losses, because individual losses often occur in ranges that fall below deductibles.

Our results indicate that catastrophe obligation guarantees would ultimately translate to a relatively small reduction in federal disaster-assistance expenditures for earthquake losses in California. At the same time, the cost to the federal government of providing catastrophe obligation guarantees is anticipated to be quite low as well. The primary cost would be repaying loans taken out by state catastrophe-insurance programs in the very low likelihood that they default. The Congressional Budget Office (2010) predicts that catastrophe obligation guarantees would cost the federal government $23.5 million per year over the next five years. Further, two factors indicate that nearly all of this cost (probably more than 90 percent) is expected to be in support of state programs other than the CEA. One is that the catastrophe-insurance

Table 8
Changes with Catastrophe Obligation Guarantees per $10 Billion in Total Earthquake Losses

Conditions	Change in Residential ITV		Change in Uninsured Residential Loss		Change in Federal Disaster Assistance[a]	
	Change	%	$ millions	%	$ millions	%
All earthquakes	0.009	8.3	−14.7	−0.32	−3.45	−0.32
Large earthquakes	0.009	8.3	−31.5	−0.71	−7.41	−0.71

[a] Change is only for disaster assistance considered in this study (F_{SBA} and F_{tax}).

market is dominated by flood, wind, and other weather-related disasters, so earthquakes would be only a small fraction of the losses and associated loans. The other is that the CEA maintains a higher claim-paying capability than some of the other state programs and would require loans only in extraordinarily large disasters. With catastrophe obligation guarantees in place, the CEA is expected to be able to suffer losses up to about $7 billion before needing loans, which is more than sufficient to cover claims in an earthquake with a total loss of well over $100 billion (Table 6).

If we assume that the annual cost to the federal government of catastrophe obligation guarantees in California is $2.3 million, our analysis suggests that the annual savings in terms of reduced disaster assistance would exceed the cost if the annual expected total loss from earthquakes was $7 billion or greater. This is equivalent to an earthquake like the simulated San Andreas Fault earthquake in Table 6 occurring every 17 years.

It may seem surprising that catastrophe obligation guarantees would allow the CEA to reduce costs by about $180 million per year[1] but are anticipated to cost the federal government only a few million dollars per year. Much of this apparent discrepancy can be understood by noting that the effect of catastrophe obligation guarantees is to allow the CEA to trade reinsurance for loans. Loan guarantees cost much less than reinsurance for access to a given amount of claim-paying capability, but they also must be repaid. Hence, the missing element is the cost to future CEA policyholders, who will have to pay increased premiums to pay off the loans after an earthquake in which the CEA had to take out loans to pay claims. Accounting for that cost would largely reconcile the apparent discrepancy. Repayment can be spread over 30 years, however, so the ultimate cost under catastrophe obligation guarantees, including that to future policyholders, is still much less than the reinsurance cost without catastrophe obligation guarantees.

While the benefit and cost to the federal government of catastrophe obligation guarantees in California are modest, the situation for other state catastrophe-insurance programs may be different. Because annual losses from weather-related disasters are much greater than earthquake losses, the benefits might be much greater in other states with catastrophe-insurance programs that would qualify for catastrophe obligation guarantees, such as Florida, Louisiana, or Texas. Analogous benefit analyses for all eligible state catastrophe-insurance programs are therefore needed before the overall net benefit to the federal government of catastrophe obligation guarantees can be estimated.

Broader Considerations for Increasing Catastrophe-Insurance Coverage

Our analysis has focused on the potential benefit of catastrophe obligation guarantees in terms of reducing direct federal disaster assistance for earthquakes in California. Our results show that the effect is small and thus that changes in insurance coverage would have to be dramatic to have an appreciable impact on uninsured loss and disaster assistance. We conclude by noting that there are other ways to increase catastrophe-insurance coverage and that increasing catastrophe-insurance coverage provides benefits beyond reducing disaster-assistance expenditures.

---

[1] The CEA collects about $600 million per year in premiums, so a 30-percent price decrease is $180 million per year.

Our analysis focused on the effect of price on earthquake-insurance coverage because price is known to be a key factor in decisions about purchasing earthquake insurance. In addition to reducing prices, coverage might also be increased through increased public education and marketing focusing on the consequences of being uninsured in an earthquake. Another avenue for potentially increasing coverage is to consider new earthquake-insurance products that provide more-attractive options for consumers. The current policy structure with high deductibles undermines the benefit of increased take-up on reducing uninsured loss and disaster assistance. When deductibles are high, increases in take-up may only modestly diminish uninsured losses, because individual losses often occur in ranges that fall below deductibles. Further, disaster assistance provided for uninsured losses that fall below deductibles would not be reduced by increased take-up.

This suggests that policy options with lower deductibles might warrant further examination. Because losses below deductibles are more likely than losses above them, reducing deductibles is more expensive per dollar of coverage than adding new policies with the current high-deductible structure. The increased cost of lower deductibles could be mitigated by reducing policy limits. Determining the optimal combination of deductible and policy limit is a complex undertaking that would require models of the expected distribution over time of losses to a property, as well as an understanding of the long-term, indirect trade-offs between regimes featuring relatively frequent, small losses (as would occur when deductibles and policy limits are both high) versus relatively infrequent, large losses (as would occur when deductibles and policy limits are both low).

Increasing the fraction of earthquake losses that are compensated by insurance may have benefits beyond reducing federal disaster-assistance expenditures. Uncompensated disaster losses might have far-reaching and sustained economic impacts on families and communities. Examples of such indirect losses include depletion of individual savings, losses to lenders from widespread defaulting of home mortgages,[2] local decreases in property values and property tax revenue, increased unemployment, decreased income tax revenue, and lower business investment and entrepreneurship. Few of these impacts would be compensated by disaster-assistance programs, so they would be reduced only by increased insurance coverage. In addition, insurance provides the ability to create incentives to reduce disaster risks. Insurers often provide premium discounts for mitigative actions, such as bolting houses to foundations and reinforcing cripple walls. The reduction in losses from such actions represents a benefit that would not be realized through disaster assistance.

---

[2] Mortgage defaults stemming from earthquake losses might represent a substantial risk to the federal government. Through the Federal Housing Administration, Federal National Mortgage Association (Fannie Mae), Federal Home Loan Mortgage Corporation (Freddie Mac), and Government National Mortgage Association (Ginnie Mae), the federal government guarantees or owns more than 90 percent of all new home mortgage originations from 2008–2010 (Federal Housing Finance Agency, 2010). This high degree of involvement in the housing market means that, if damage from an earthquake were to cause a large number homeowners to default on their mortgages, the federal government could end up with a substantial liability.

Appendix

This appendix presents the calculations used to derive Equations 1 and 6 in the main text.

Equation 1

The price elasticity of demand, E, is defined as the fractional change in demand divided by the fractional change in price:

$$E = \frac{\left(demand_{final} - demand_{initial}\right)/demand_{initial}}{\left(price_{final} - price_{initial}\right)/price_{initial}}.$$

(A.1)

We express demand for earthquake insurance in terms of the ITV, which is defined as the total residential structural coverage (sum of coverage A policy limits minus sum of policy deductibles) for earthquake insurance divided by the total residential housing structural value. When comparing the ITV for earthquake insurance with catastrophe obligation guarantees (COG) to current conditions (o), we get

$$E = \frac{\left(ITV_{COG} - ITV_o\right)/ITV_o}{\Delta p/p_o},$$

(A.2)

where p is price and $\Delta p = \left(p_{COG} - p_o\right)$. Distinguishing the pricing of CEA insurance from that of private-market (non-CEA) insurance and noting that the CEA writes 61 percent of the residential earthquake-insurance coverage in California, we can write,

$$E = \frac{\left(ITV_{COG} - ITV_o\right)/ITV_o}{0.61\left(\Delta p/p_o\right)_{CEA} + 0.39\left(\Delta p/p_o\right)_{priv}},$$

(A.3)

where the subscript *priv* refers to private-market policies.

Since private-market insurers are not eligible for catastrophe obligation guarantees, only the CEA prices will directly change in response. While it is possible that lower CEA pricing would cause a price reaction on the part of private-market participants, we have no clear picture of how the pricing of CEA and private-market insurers are related. We therefore assume that the price of private-market insurance will not change, in which case Equation A.3 reduces to

$$E = \frac{\left(ITV_{COG} - ITV_o\right)/ITV_o}{0.61\left(\Delta p / p_o\right)_{CEA}}.$$

(A.4)

Rearranging Equation A.4 gives Equation 1 in the main text:

$$ITV_{COG} = ITV_o\left(1 + 0.61E\left(\frac{\Delta p}{p_o}\right)_{CEA}\right).$$

(1)

Regardless of whether or not private-market prices change, it is likely that some of the CEA demand changes would come at the expense of the CEA's private-market competition. In that case, the net increase in earthquake coverage in California would be less than the increase in CEA business. This is a potentially important effect, but we have no way to estimate its magnitude.

Equation 6

Equation 5 can be rewritten as

$$L_R^I = L_R R_I ITV.$$

(A.5)

The uninsured residential loss is the difference between the total residential loss and residential loss reimbursed by insurance:

$$L_R^U = L_R - L_R^I.$$

(A.6)

Substituting Equation A.5 into Equation A.6 gives Equation 6 in the main text:

$$L_R^U = L_R\left(1 - R_I ITV\right).$$

(6)

References

Barro, Robert J., and Charles J. Redlick, *Macroeconomic Effects from Government Purchases and Taxes*, Cambridge, Mass.: National Bureau of Economic Research, working paper 15369, September 2009. As of September 7, 2010:
http://papers.nber.org/papers/15369

Bea, Keith, *Federal Stafford Act Disaster Assistance: Presidential Declarations, Eligible Activities, and Funding*, Washington, D.C.: Congressional Research Service, Library of Congress, RL33053, March 16, 2010.

Browne, Mark J., and Robert E. Hoyt, "The Demand for Flood Insurance: Empirical Evidence," *Journal of Risk and Uncertainty*, Vol. 20, No. 3, 2000, pp. 291–306.

California Department of Insurance, "Earthquake Premium and Policy Count Data Call Summary of 2009 Residential and Commercial Market Totals," May 12, 2010. As of September 7, 2010:
http://www.insurance.ca.gov/0400-news/0200-studies reports/0300-earthquake-study/upload/EQ2009SummaryMay1210.pdf

Congressional Budget Office, *Cost Estimate: H.R. 2555 Homeowners' Defense Act of 2010*, Washington, D.C., June 2, 2010. As of September 7, 2010:
http://www.cbo.gov/ftpdocs/115xx/doc11548/hr2555.pdf

Cummins, J. David, Michael Suher, and George Zanjani, "Federal Financial Exposure to Natural Catastrophe Risk," in Deborah Lucas, ed., *Measuring and Managing Federal Financial Risk*, Chicago and London: University of Chicago Press, National Bureau of Economic Research conference report, 2010, pp. 61–96.

Dixon, Lloyd, Noreen Clancy, Seth A. Seabury, and Adrian Overton, *The National Flood Insurance Program's Market Penetration Rate: Estimates and Policy Implications*, Santa Monica, Calif.: RAND Corporation, TR-300-FEMA, 2006. As of September 7, 2010:
http://www.rand.org/pubs/technical_reports/TR300/

Eguchi, Ronald T., James D. Goltz, Craig E. Taylor, Stephanie E. Chang, Paul J. Flores, Laurie A. Johnson, Hope A. Seligson, and Neil C. Blais, "Direct Economic Losses in the Northridge Earthquake: A Three Year Post-Event Perspective," *Earthquake Spectra*, Vol. 14, No. 2, 1998, pp. 245–264. As of September 7, 2010:
http://authors.library.caltech.edu/7079/

Federal Emergency Management Agency, *Robert T. Stafford Disaster Relief and Emergency Assistance Act, P.L. 93-288 as Amended*, Washington, D.C., FEMA 592, August 2007.

———, *Help After a Disaster: Applicant's Guide to the Individuals and Households Program*, Washington, D.C., FEMA 545, July 2008. As of September 7, 2010:
http://purl.access.gpo.gov/GPO/LPS113786

———, *National Disaster Housing Strategy*, Washington, D.C., January 16, 2009. As of September 7, 2010:
http://purl.access.gpo.gov/GPO/LPS107999

Federal Housing Finance Agency, *Conservator's Report on the Enterprises' Financial Performance Second Quarter 2010*, August 26, 2010. As of September 7, 2010:
http://www.fhfa.gov/webfiles/16591/ConservatorsRpt82610.pdf

FEMA—*See* Federal Emergency Management Agency.

GAO—*See* U.S. General Accounting Office (before 2004) or U.S. Government Accountability Office (since 2004).

Grace, Martin F., Robert W. Klein, and Paul R. Kleindorfer, "Homeowner's Insurance with Bundled Catastrophe Coverage," *Journal of Risk and Insurance*, Vol. 71, No. 3, September 2004, pp. 351–379.

Grossi, Patricia, and Mary Lou Zoback, *Catastrophe Modeling and California Earthquake Risk: A 20-Year Perspective*, Risk Management Solutions special report, 2009. As of September 7, 2010:
http://www.rms.com/publications/LomaPrieta_20Years.pdf

H.R.2555—*See* U.S. House of Representatives (2009a).

H.R.4014—*See* U.S. House of Representatives (2009b).

Hung, Hung-Chih, "The Attitude Towards Flood Insurance Purchase When Respondents' Preferences Are Uncertain: A Fuzzy Approach," *Journal of Risk Research*, Vol. 12, No. 2, 2009, pp. 239–258.

Internal Revenue Service, "Katrina Emergency Tax Relief Act of 2005," last updated November 1, 2007. As of September 7, 2010:
http://www.irs.gov/newsroom/article/0,,id=149391,00.html

———, *Casualties, Disasters, and Thefts: For Use in Preparing 2009 Returns*, Washington, D.C., publication 547, 2009. As of September 7, 2010:
http://www.irs.gov/publications/p547/index.html

IRS—*See* Internal Revenue Service.

Kriesel, Warren, and Craig Landry, "Participation in the National Flood Insurance Program: An Empirical Analysis for Coastal Properties," *Journal of Risk and Insurance*, Vol. 71, No. 3, September 2004, pp. 405–420.

Lott, Neal, and Tom Ross, "Tracking and Evaluating U.S. Billion Dollar Weather Disasters, 1980–2005," Asheville, N.C.: U.S. National Oceanographic and Atmospheric Administration National Climatic Data Center, c. 2006. As of September 7, 2010:
http://www1.ncdc.noaa.gov/pub/data/papers/200686ams1.2nlfree.pdf

Marshall, Dan, general counsel, California Earthquake Authority, "Earthquake Insurance for Homeowners," presentation at the Bay Area Earthquake Alliance Meeting, July 29, 2009. As of September 7, 2010:
http://bayquakealliance.org/meetings/20090729/

Petak, William J., and Shirin Elahi, *The Northridge Earthquake, USA and its Economic and Social Impacts*, presented at the EuroConference on Global Change and Catastrophe Risk Management Earthquake Risks in Europe, International Institute for Applied Systems Analysis, Laxenburg, Austria, July 6–9, 2000. As of September 7, 2010:
http://www.iiasa.ac.at/Research/RMP/july2000/Papers/Northridge_0401.pdf

Pomeroy, Glenn, chief executive officer, California Earthquake Authority, testimony before the U.S. House of Representatives Committee on Financial Services Subcommittee on Housing and Community Opportunity and Subcommittee on Capital Markets, Insurance, and Government-Sponsored Enterprises hearing, "Approaches to Mitigating and Managing Natural Catastrophe Risk: H.R.2555, The Homeowners' Defense Act," March 10, 2010. As of September 7, 2010:
http://www.house.gov/apps/list/hearing/financialsvcs_dem/pomeroy.pdf

Public Law 93-288, Robert T. Stafford Disaster Relief and Emergency Assistance Act, as amended by Public Law 100-707, November 23, 1988.

Public Law 107-134, Victims of Terrorism Tax Relief Act of 2001, January 23, 2002. As of September 7, 2010:
http://thomas.loc.gov/cgi-bin/bdquery/z?d107:h.r.02884:

Public Law 109-73, Katrina Emergency Tax Relief Act of 2005, September 23, 2005. As of September 7, 2010:
http://thomas.loc.gov/cgi-bin/bdquery/z?d109:HR03768:

Public Law 109-135, Gulf Opportunity Zone Act of 2005, December 21, 2005. As of September 7, 2010:
http://thomas.loc.gov/cgi-bin/bdquery/z?d109:HR04440:

S.886—*See* U.S. Senate (2009).

U.S. General Accounting Office, *The Effect of Premium Increases on Achieving the National Flood Insurance Program's Objectives: Report*, Washington, D.C., GAO/RCED-83-107, February 28, 1983.

U.S. Government Accountability Office, *National Flood Insurance Program: Financial Challenges Underscore Need for Improved Oversight of Mitigation Programs and Key Contracts*, Washington, D.C., GAO-08-437, June 2008. As of September 7, 2010:
http://purl.access.gpo.gov/GPO/LPS111728

U.S. House of Representatives, Homeowners' Defense Act of 2010, H.R.2555, introduced May 21, 2009a; placed on union calendar July 13, 2010. As of September 7, 2010:
http://thomas.loc.gov/cgi-bin/bdquery/z?d111:h.r.02555:

————, Catastrophe Obligation Guarantee Act of 2009, H.R.4014 introduced and referred to the Committee on Financial Services, November 4, 2009. As of September 7, 2010:
http://thomas.loc.gov/cgi-bin/bdquery/z?d111:HR04014:

U.S. Senate, Catastrophe Obligation Guarantee Act, S.886, introduced, read, and referred to the Committee on Banking, Housing, and Urban Affairs, April 23, 2009. As of September 7, 2010:
http://thomas.loc.gov/cgi-bin/bdquery/z?d111:SN00886:

Wolfe, M. Ann, *Homeland Security: 9/11 Victim Relief Funds*, Washington, D.C.: Congressional Research Service, Library of Congress, 03-RL-31716, updated March 27, 2003.